Collins

Maths
in 5 minutes

Quick practice activities

Chief editor: Zhou Jieying
Consultant: Fan Lianghuo

CONTENTS

HOW TO USE THIS BOOK

The best way to help your child to build their confidence in maths and improve their number skills is to give them lots and lots of practice in the key facts and skills.

Written by maths experts, this series will help your child to become fluent in number facts, and help them to recall them quickly – both are essential for succeeding in maths.

This book provides ready-to-practise questions that comprehensively cover the number curriculum for Year 1. It contains 40 topic-based tests, each 5 minutes long, to help your child build up their mathematical fluency day-by-day.

Each test is divided into three Steps:

- **Step 1: Warm-up (1 minute)**
  This exercise helps your child to revise maths they should already know and gives them preparation for Step 2.

- **Step 2: Rapid calculation (2½ minutes)**
  This exercise is a set of questions focused on the topic area being tested.

- **Step 3: Challenge (1½ minutes)**
  This is a more testing exercise designed to stretch your child's mental abilities.

Some of the tests also include:

- a Tip to help your child answer questions of a particular type.

- a Mind Gym puzzle – this is a further test of mental agility and is not included in the 5-minute time allocation.

Your child should attempt to answer as many questions as possible in the time allowed at each Step. Answers are provided at the back of the book.

To help to measure progress, each test includes boxes for recording the date of the test, the total score obtained and the total time taken.

ACKNOWLEDGEMENTS

The authors and publisher are grateful to the copyright holders for permission to use quoted materials and images.

All images are © HarperCollinsPublishers Ltd and © Shutterstock.com

Every effort has been made to trace copyright holders and obtain their permission for the use of copyright material. The authors and publisher will gladly receive information enabling them to rectify any error or omission in subsequent editions. All facts are correct at time of going to press.

Published by Collins in association with East China Normal University Press

Collins
An imprint of HarperCollinsPublishers
1 London Bridge Street
London SE1 9GF

HarperCollinsPublishers
Macken House
39/40 Mayor Street Upper
Dublin 1
D01 C9W8
Ireland

ISBN: 978-0-00-831108-7

First published 2019
This edition published 2020
Previously published as Letts

10 9 8 7 6

©HarperCollinsPublishers Ltd. 2020
©East China Normal University Press Ltd.,
©Zhou Jieying

British Library Cataloguing in Publication Data.

A CIP record of this book is available from the British Library.

Publisher: Fiona McGlade
Consultant: Fan Lianghuo
Authors: Zhou Jieying, Xu Jing and Yang Chenmin
Editors: Ni Ming and Xu Huiping
Contributor: Paul Hodge
Project Management and Editorial: Richard Toms, Lauren Murray and Marie Taylor
Cover Design: Sarah Duxbury and Kevin Robbins
Inside Concept Design: Paul Oates and Ian Wrigley
Layout: Jouve India Private Limited
Printed and bound in the UK

MIX
Paper | Supporting responsible forestry
FSC www.fsc.org FSC™ C007454

This book contains FSC™ certified paper and other controlled sources to ensure responsible forest management.

For more information visit: www.harpercollins.co.uk/green

Date: _____

Day of Week: _____

STEP 1 (1 min) **Warm-up**

Start the timer

Write the number of pieces of fruit in each group.

1.  ☐

2.  ☐

3.  ☐

4. ☐

5. ☐

6.  ☐

7. ☐

8. ☐

9.  ☐

STEP 2 (2.5 min) **Rapid calculation**

Start the timer

Draw triangles (Δ) to match the number.

1. 5

2. 3

3. 4

4. 2

Draw a circle around sets of the same fruit. The first one has been done for you.

5.

6.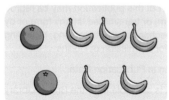

7.

8.

9.

10.

STEP 3 (1.5 min) **Challenge**

Start the timer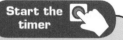

Sort by colour, shape and size, and then count.

Type	☆	★	◇	◇	● + ○
Total					

Time spent: _____ min _____ sec. Total: _____ out of 23

STEP 1 (1 min) Warm-up

Start the timer

Write the number of pieces of fruit in each group.

1. ☐

2.  ☐

3. ☐

4. ☐

5. ☐

6. ☐

7. ☐

8. ☐

9. ☐

STEP 2 (2.5 min) Rapid calculation

Start the timer

Draw triangles (△) to match the number.

1. 7

2. 6

3. 8

4. 10

Draw a circle around sets of the same fruit.

5.

6.

7.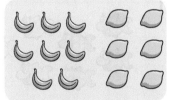

8.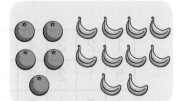

9.

10.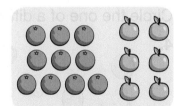

STEP 3 (1.5 min) Challenge

Start the timer

Sort by colour, shape and size, and then count.

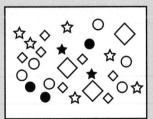

Type	☆ + ★	◇ + ◇	● + ○	★ + ●	★ + ◇
Total					

Date: _____

Day of Week: _____

Start the timer

Match each group of objects to its number. One has been done for you.

| 1 | 2 | 3 | 4 | 5 | ⑥ | 7 | 8 | 9 | 10 |

Ten   Seven   ⬭Six⬭   One   Four   Two   Five   Nine   Three   Eight

Start the timer

Sort and circle.

1.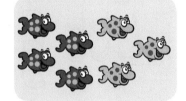

2.

3.

Circle the one of a different kind.

4.

5.

1	2	3	4	5
5	1	2	3	4
4	5	1	2	3

6. Find the pattern and complete the table so that 1, 2, 3, 4 and 5 are in each row and column.

Start the timer

Write down three ways of sorting these shapes.

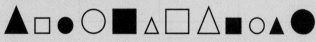

Sorted by: _____

Sorted by: _____

Sorted by: _____

Time spent: _____ min _____ sec. Total: _____ out of 18

©HarperCollins*Publishers* 2019

Let's count (2)   4

STEP 1 Warm-up

Start the timer

Match each group of objects to its number.

1	2	3	4	5	6	7	8	9	10

Two	Eight	Five	Nine	Ten	Seven	One	Four	Six	Three

STEP 2 Rapid calculation

Start the timer

Complete the table to show how many of each. The first row has been done for you.

Objects	Dots	Word	Number
🧸🧸🧸🧸	●●●●○	Four	4
🍎🍎	○○○○○		
⛵⛵⛵⛵⛵⛵⛵	○○○○○ ○○○○○		
⚫⚫⚫	○○○○○		
🍌🍌🍌🍌🍌🍌🍌	○○○○○ ○○○○○		
🚂🚂🚂🚂🚂	○○○○○		
🧱🧱🧱🧱🧱🧱🧱🧱🧱	○○○○○ ○○○○○		
🍓🍓🍓🍓🍓🍓🍓🍓	○○○○○ ○○○○○		

STEP 3  Challenge

Start the timer

Find the pattern and draw the next group of shapes.

○△○△○△ _____

○○△△○△ _____

○△○○△○△○○△ _____

5 Let's count (3)

STEP 1 (1 min) Warm-up

Start the timer

Count and colour. The first one has been done for you.

| 5 | ●●●●● ○○○○○ | 8 | ○○○○○ ○○○○○ | 6 | ○○○○○ ○○○○○ |
| 9 | ○○○○○ ○○○○○ | 7 | ○○○○○ ○○○○○ | 10 | ○○○○○ ○○○○○ |

STEP 2 (2.5 min) Rapid calculation

Start the timer

Circle the one of a different kind.

1.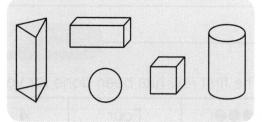

2.

3. Put in the missing numbers or shapes. The first one has been done for you.

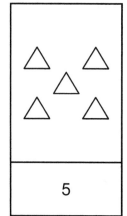 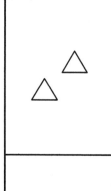 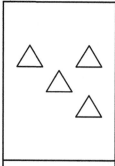 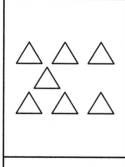 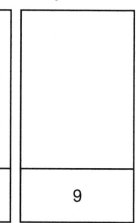

| 5 | | | | 9 |

STEP 3 (1.5 min) Challenge

Start the timer

Match each object to a box in which it will fit.

Objects:

     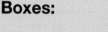 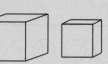

Boxes:

Time spent: _____ min _____ sec. Total: _____ out of 16

STEP 1 (1 min) Warm-up

Start the timer

Write numbers in the boxes to show the order of these animals. Two have been done.

Number: | 1 | | 3 | | | | | | |

STEP 2 (2.5 min) Rapid calculation

Start the timer

1.

From **left** to **right**:

In the 2nd box there are ☐ keys; in the 4th box there are ☐ keys; and in the 3rd

box, there are ☐ keys. In the ☐ box, there are 3 keys; in the ☐ box,

there are 6 keys; and in the ☐ box, there are 8 keys.

2.

From **right** to **left**:

In the 1st box, there are ☐ pencils; in the 4th box, there are ☐ pencils; and in

the 6th box, there are ☐ pencils. In the ☐ box, there are 4 pencils; in the

☐ box, there are 9 pencils; and in the ☐ box, there are 7 pencils.

STEP 3 (1.5 min) Challenge

Start the timer

Draw crayons to make the pencil the 6th object when you count from **right** to **left**.

 ☐

Draw stars to make the moon the 7th object when you count from **right** to **left**.

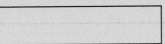 ☐

©HarperCollins*Publishers* 2019

Time spent: _____ min _____ sec. Total: _____ out of 21

9

7 Let's compare

STEP 1 (1 min) Warm-up

Start the timer

Fill in the missing numbers in these number lines.

| 1 | | 3 |

| 4 | | | 7 |

| 10 | | 8 | | | 5 | | | | | 0 |

| 8 | | 6 |

| | | 6 | 7 | |

| 0 | | | 3 | | | | | | 9 | |

STEP 2 (2.5 min) Rapid calculation

Start the timer

Choose a number to complete each statement.

1. 6 is greater than ☐   **2.** ☐ is less than 8   **3.** ☐ is greater than 2

4. 5 is less than ☐   **5.** 9 is greater than ☐   **6.** ☐ is greater than 1

Arrange each set of numbers below from **smallest** to **largest**.

7. 3, 5, 9, 2, 8 _____

8. 7, 3, 9, 10, 5 _____

9. 3, 0, 5, 7, 9 _____

Arrange each set of numbers below from **largest** to **smallest**.

10. 4, 8, 10, 7, 1 _____

11. 5, 0, 4, 9, 2 _____

12. 8, 5, 1, 6, 3 _____

STEP 3 (1.5 min) Challenge

Start the timer

1. Write three numbers less than 7. _____

2. Write three numbers greater than 7. _____

3. Write three numbers less than 5. _____

4. Write three numbers greater than 5. _____

5. Write three numbers less than 3. _____

6. Write three numbers greater than 3. _____

Time spent: _____ min _____ sec. Total: _____ out of 24

Date: _____

Day of Week: _____

©HarperCollins*Publishers* 2019

STEP 1 (1 min) Warm-up

Start the timer

Count the shapes and fill in the blanks. The first one has been done for you.

5	5	5	5
○○●●●	●●○●●	○●●○○	○●○○○
○ **2** ● **3**	○ ___ ● ___	○ ___ ● ___	○ ___ ● ___

6	6	6	6	6
○○○●●●	●●○●○●	○○○●●○○	○○●●○○●	●●●○●●
○ ___ ● ___	○ ___ ● ___	○ ___ ● ___	○ ___ ● ___	○ ___ ● ___

STEP 2 (2.5 min) Rapid calculation

Start the timer

Count the shapes and fill in the blanks.

●●●○○●	○ ___ ● ___	○○○●●●○	○ ___ ● ___
●●○○○○	○ ___ ● ___	○○●●●●	○ ___ ● ___
●●○○○●	○ ___ ● ___	○○○○●●	○ ___ ● ___
●○○○●○	○ ___ ● ___	●○○●●○	○ ___ ● ___
●○○○●●	○ ___ ● ___	●●●○○○●	○ ___ ● ___
○●○○○●	○ ___ ● ___	○○●●○○	○ ___ ● ___
○●●●●●	○ ___ ● ___	○○●○●○●	○ ___ ● ___
●○○●○●	○ ___ ● ___	○○●○●●●	○ ___ ● ___
●●○●●	○ ___ ● ___	○●●○○○	○ ___ ● ___
○●●○●	○ ___ ● ___	○●●○●●	○ ___ ● ___

STEP 3 (1.5 min) Challenge

Start the timer

Complete the number bonds.

5: 1, □   5: 3, □   5: □, 4   5: □, 2   5: 0, □

6: 2, □   6: 4, □   6: □, 3   6: □, 5   6: 1, □

Date: _____

Day of Week: _____

STEP 1 (1 min) **Warm-up**

Start the timer

Count the shapes and fill in the blanks.

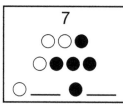
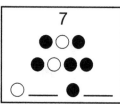
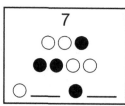
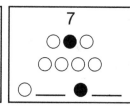
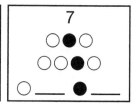

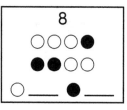
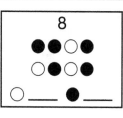
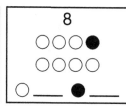
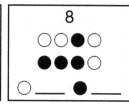
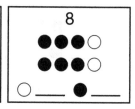

STEP 2 (2.5 min) **Rapid calculation**

Start the timer

Count the shapes and fill in the blanks.

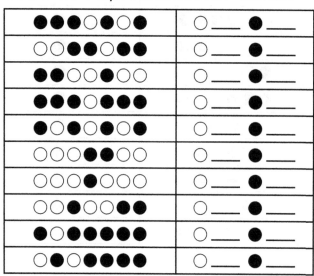

STEP 3 (1.5 min) **Challenge**

Start the timer

Complete the number bonds.

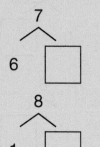
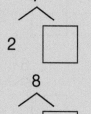
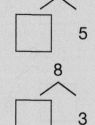
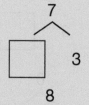

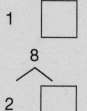

Time spent: _____ min _____ sec. Total: _____ out of 40

Date: _____

Day of Week: _____

Start the timer

STEP 1 (1 min) Warm-up

Count and fill in the blanks.

9	9	9	9	9
○○●● ●○○●● ○ ___ ● ___	●○○● ●○○●● ○ ___ ● ___	○●●● ●●●●● ○ ___ ● ___	○●●○ ○○●●○ ○ ___ ● ___	○●○○ ○○●○○ ○ ___ ● ___

10	10	10	10	10
○○○●○ ●●○○○ ○ ___ ● ___	●●○●○ ○●○●● ○ ___ ● ___	○○○●○ ○○○○○ ○ ___ ● ___	○○●○○ ●○○○○ ○ ___ ● ___	●●●○○ ●●●○○ ○ ___ ● ___

Start the timer

STEP 2 (2.5 min) Rapid calculation

Count and fill in the blanks.

●●●○○○○●○○	○ ___ ● ___
○○●●○●●●●	○ ___ ● ___
●●●○●●●●●	○ ___ ● ___
●●●○○●●●●	○ ___ ● ___
●○●○●○●○●	○ ___ ● ___
○○○●●○○○○	○ ___ ● ___
○○○●●●○○○	○ ___ ● ___
○○○○●○○○○	○ ___ ● ___
●○○○○○●●●	○ ___ ● ___
○●○●○○○○○	○ ___ ● ___

○○●●●○●●○○	○ ___ ● ___
●●○○●○○●●●	○ ___ ● ___
○○●○○●○○○○	○ ___ ● ___
●○●○○●●●●●	○ ___ ● ___
●●●●●○●●●	○ ___ ● ___
●○○●○●●○○●	○ ___ ● ___
●●●●●●●●○	○ ___ ● ___
○○○○●●●●○	○ ___ ● ___
○○●○○○○○○●	○ ___ ● ___
●○●●●●○○●●	○ ___ ● ___

Start the timer

STEP 3 (1.5 min) Challenge

Complete the number bonds.

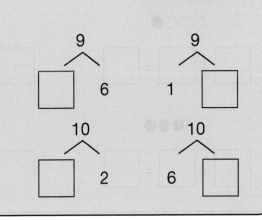

9 ∧ 4 [ ]   9 ∧ 5 [ ]   9 ∧ [ ] 3   9 ∧ [ ] 6   9 ∧ 1 [ ]

10 ∧ 5 [ ]   10 ∧ 1 [ ]   10 ∧ [ ] 7   10 ∧ [ ] 2   10 ∧ 6 [ ]

11 Addition (1)

Date: _____

Day of Week: _____

STEP 1 (1 min) **Warm-up**

Start the timer

Answer these.

1 + 1 = ☐   1 + 4 = ☐   4 + 1 = ☐   2 + 2 = ☐   1 + 3 = ☐

1 + 2 = ☐   2 + 1 = ☐   2 + 3 = ☐   3 + 1 = ☐   3 + 2 = ☐

STEP 2 (2.5 min) **Rapid calculation**

Start the timer

Answer these.

2 + 2 = ☐   2 + 0 = ☐   2 + 3 = ☐   0 + 1 = ☐   1 + 3 = ☐

0 + 5 = ☐   4 + 0 = ☐   1 + 2 = ☐   4 + 1 = ☐   3 + 2 = ☐

2 + 1 = ☐   1 + 4 = ☐   1 + 1 = ☐   0 + 2 = ☐   3 + 0 = ☐

4 + 0 = ☐   3 + 1 = ☐   0 + 3 = ☐   4 + 1 = ☐   5 + 0 = ☐

STEP 3 (1.5 min) **Challenge**

Start the timer

Write the addition sentence for each picture.

○○●   ○○○○●   ○●●●●

☐ + ☐ = ☐   ☐ + ☐ = ☐   ☐ + ☐ = ☐

○○○●   ○○○●●   ○●●●

☐ + ☐ = ☐   ☐ + ☐ = ☐   ☐ + ☐ = ☐

○○●●●   ○●●

☐ + ☐ = ☐   ☐ + ☐ = ☐

Time spent: _____ min _____ sec. Total: _____ out of 38

Date: _____

Day of Week: _____

STEP 1 (1 min) Warm-up

Start the timer

Answer these.

4 + 2 = ☐   3 + 0 = ☐   5 + 1 = ☐   3 + 6 = ☐   2 + 6 = ☐

3 + 3 = ☐   2 + 2 = ☐   7 + 0 = ☐   1 + 2 = ☐   6 + 1 = ☐

STEP 2 (2.5 min) Rapid calculation

Start the timer

Answer these.

5 + 5 = ☐   1 + 7 = ☐   2 + 5 = ☐   9 + 1 = ☐   1 + 0 = ☐

7 + 2 = ☐   3 + 1 = ☐   3 + 5 = ☐   6 + 4 = ☐   5 + 4 = ☐

1 + 6 = ☐   3 + 4 = ☐   8 + 0 = ☐   5 + 2 = ☐   1 + 4 = ☐

6 + 3 = ☐   4 + 1 = ☐   1 + 9 = ☐   0 + 3 = ☐   4 + 5 = ☐

STEP 3 (1.5 min) Challenge

Start the timer

Write the addition sentence for each picture.

☐ + ☐ = ☐   ☐ + ☐ = ☐   ☐ + ☐ = ☐

☐ + ☐ = ☐   ☐ + ☐ = ☐   ☐ + ☐ = ☐

☐ + ☐ = ☐   ☐ + ☐ = ☐   ☐ + ☐ = ☐

13 Addition (3)

STEP 1 (1 min) Warm-up

Start the timer

Answer these.

8 + 2 = ☐   7 + 1 = ☐   5 + 0 = ☐   9 + 1 = ☐   5 + 3 = ☐

1 + 4 = ☐   4 + 5 = ☐   8 + 1 = ☐   0 + 9 = ☐   4 + 6 = ☐

STEP 2 (2.5 min) Rapid calculation

Start the timer

Answer these.

2 + 6 = ☐   3 + 5 = ☐   7 + 3 = ☐   4 + 3 = ☐   5 + 4 = ☐

4 + 0 = ☐   5 + 1 = ☐   0 + 2 = ☐   2 + 2 = ☐   1 + 5 = ☐

3 + 1 = ☐   2 + 7 = ☐   1 + 8 = ☐   3 + 7 = ☐   6 + 1 = ☐

1 + 7 = ☐   2 + 1 = ☐   3 + 0 = ☐   2 + 5 = ☐   1 + 6 = ☐

STEP 3 (1.5 min) Challenge

Start the timer

Write the addition sentence for each picture.

○○○
●●●●●●
☐ + ☐ = ☐

○○○○○○
●●
☐ + ☐ = ☐

○○
●●●
☐ + ☐ = ☐

○○○○
●●
☐ + ☐ = ☐

○○○○
●●●●
☐ + ☐ = ☐

○○
●●●●●●●●
☐ + ☐ = ☐

○○○○○○
●●●●
☐ + ☐ = ☐

○○○○○○
●●●
☐ + ☐ = ☐

Time spent: _____ min _____ sec. Total: _____ out of 38

©HarperCollinsPublishers 2019

Date: _____

Day of Week: _____

STEP 1 (1 min) Warm-up

Start the timer

Answer these.

4 + 4 = ☐   5 + 1 = ☐   8 + 2 = ☐   0 + 7 = ☐   1 + 5 = ☐

1 + 6 = ☐   6 + 2 = ☐   4 + 5 = ☐   3 + 2 = ☐   3 + 1 = ☐

STEP 2 (2.5 min) Rapid calculation

Start the timer

Answer these.

5 + 0 = ☐   8 + 1 = ☐   2 + 1 = ☐   7 + 1 = ☐   7 + 2 = ☐

1 + 1 = ☐   2 + 4 = ☐   5 + 4 = ☐   0 + 3 = ☐   4 + 3 = ☐

6 + 0 = ☐   3 + 3 = ☐   3 + 7 = ☐   4 + 6 = ☐   6 + 3 = ☐

2 + 2 = ☐   4 + 1 = ☐   2 + 5 = ☐   2 + 6 = ☐   5 + 5 = ☐

STEP 3 (1.5 min) Challenge

Start the timer

Write the addition sentence for each picture.

○○○○
●●

☐ + ☐ = ☐

○○
●●●●●●●

☐ + ☐ = ☐

○○
●●●●●●●●

☐ + ☐ = ☐

○○○
●●●●●

☐ + ☐ = ☐

○○○○○○○○○
●

☐ + ☐ = ☐

○○○○○○
●●●●

☐ + ☐ = ☐

○○
●●●

☐ + ☐ = ☐

○
●●●●●●●●●

☐ + ☐ = ☐

Time spent: _____ min _____ sec. Total: _____ out of 38

Date: _____

Day of Week: _____

STEP 1 ⏱ **Warm-up**

Start the timer

Answer these.

2 – 1 = ☐   3 – 2 = ☐   5 – 2 = ☐   4 – 1 = ☐   4 – 3 = ☐

3 – 1 = ☐   5 – 4 = ☐   4 – 2 = ☐   5 – 1 = ☐   5 – 3 = ☐

STEP 2 ⏱ **Rapid calculation**

Start the timer

Answer these.

5 – 2 = ☐   3 – 0 = ☐   4 – 2 = ☐   2 – 2 = ☐   3 – 1 = ☐

5 – 1 = ☐   4 – 4 = ☐   4 – 3 = ☐   2 – 1 = ☐   4 – 1 = ☐

5 – 3 = ☐   1 – 0 = ☐   5 – 4 = ☐   4 – 0 = ☐   1 – 1 = ☐

3 – 3 = ☐   3 – 2 = ☐   5 – 5 = ☐   2 – 0 = ☐   0 – 0 = ☐

STEP 3 ⏱ **Challenge**

Start the timer

Fill in the missing numbers.

2 + ☐ = 5   4 + ☐ = 5   0 + ☐ = 3

5 – ☐ = 1   4 – ☐ = 0   ☐ – 3 = 2

3 – ☐ = 2   5 – ☐ = 1   ☐ – 2 = 2

Time spent: _____ min _____ sec. Total: _____ out of 39

Date: _____

Day of Week: _____

STEP 1 (1 min) Warm-up

Start the timer

Answer these.

8 − 5 = ☐   8 − 2 = ☐   10 − 7 = ☐   9 − 4 = ☐   7 − 4 = ☐

6 − 4 = ☐   2 − 1 = ☐   4 − 2 = ☐   9 − 5 = ☐   8 − 0 = ☐

STEP 2 (2.5 min) Rapid calculation

Start the timer

Answer these.

5 − 4 = ☐   10 − 9 = ☐   9 − 3 = ☐   6 − 0 = ☐   6 − 5 = ☐

9 − 7 = ☐   7 − 3 = ☐   10 − 0 = ☐   7 − 5 = ☐   9 − 1 = ☐

7 − 1 = ☐   8 − 6 = ☐   10 − 8 = ☐   10 − 5 = ☐   10 − 3 = ☐

6 − 1 = ☐   7 − 0 = ☐   4 − 1 = ☐   5 − 1 = ☐   6 − 3 = ☐

STEP 3 (1.5 min) Challenge

Start the timer

Fill in the missing numbers.

☐ − 1 = 2   7 − ☐ = 1

☐ − 2 = 3   ☐ − 1 = 7

4 − ☐ = 1   ☐ − 2 = 8

3 − ☐ = 1   5 − ☐ = 2

Mind Gym

Five triangles are made with nine matchsticks, as shown. Cross through three matchsticks to make only one triangle.

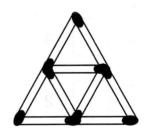

Date: _____

Day of Week: _____

Start the timer

Answer these.

$8 - 2 = \boxed{}$ $10 - 5 = \boxed{}$ $10 - 6 = \boxed{}$ $4 - 2 = \boxed{}$ $8 - 6 = \boxed{}$

$7 - 5 = \boxed{}$ $8 - 7 = \boxed{}$ $3 - 2 = \boxed{}$ $4 - 1 = \boxed{}$ $9 - 1 = \boxed{}$

STEP 2 (2.5 min) **Rapid calculation**

Start the timer

Answer these.

$9 - 7 = \boxed{}$ $10 - 3 = \boxed{}$ $8 - 0 = \boxed{}$ $8 - 5 = \boxed{}$ $6 - 4 = \boxed{}$

$9 - 6 = \boxed{}$ $7 - 1 = \boxed{}$ $9 - 8 = \boxed{}$ $5 - 4 = \boxed{}$ $10 - 2 = \boxed{}$

$6 - 3 = \boxed{}$ $3 - 1 = \boxed{}$ $10 - 8 = \boxed{}$ $7 - 6 = \boxed{}$ $6 - 2 = \boxed{}$

$8 - 4 = \boxed{}$ $8 - 3 = \boxed{}$ $6 - 5 = \boxed{}$ $10 - 0 = \boxed{}$ $7 - 3 = \boxed{}$

STEP 3 (1.5 min) **Challenge**

Start the timer

Fill in the missing numbers.

$\boxed{} - 7 = 3$ $7 - \boxed{} = 3$

$7 - \boxed{} = 0$ $\boxed{} - 9 = 1$

$5 - \boxed{} = 4$ $5 - \boxed{} = 3$

$5 - \boxed{} = 2$ $5 - \boxed{} = 1$

Mind Gym

Six triangles are made with some matchsticks, as shown. Cross through three matchsticks to leave only three triangles.

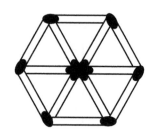

Time spent: _____ min _____ sec. Total: _____ out of 38

Date: _____

Day of Week: _____

STEP 1 (1 min) Warm-up

Start the timer

Answer these.

4 − 1 = ☐   10 − 6 = ☐   10 − 2 = ☐   8 − 6 = ☐   7 − 6 = ☐

9 − 5 = ☐   10 − 5 = ☐   10 − 4 = ☐   10 − 1 = ☐   10 − 3 = ☐

STEP 2 (2.5 min) Rapid calculation

Start the timer

Answer these.

3 − 2 = ☐   5 − 3 = ☐   5 − 2 = ☐   7 − 4 = ☐   6 − 1 = ☐

8 − 7 = ☐   9 − 6 = ☐   9 − 7 = ☐   4 − 3 = ☐   6 − 5 = ☐

8 − 4 = ☐   6 − 3 = ☐   9 − 8 = ☐   8 − 2 = ☐   5 − 1 = ☐

10 − 9 = ☐   7 − 2 = ☐   9 − 1 = ☐   4 − 2 = ☐   9 − 4 = ☐

STEP 3 (1.5 min) Challenge

Start the timer

Fill in the missing numbers.

6 − ☐ = 4   7 − ☐ = 6   ☐ − 3 = 4   ☐ − 4 = 1

☐ + 1 = 1   10 + ☐ = 10   2 + ☐ = 6   4 + ☐ = 6

☐ − 5 = 1   ☐ − 0 = 3   ☐ − 6 = 2   ☐ − 7 = 2

Time spent: _____ min _____ sec. Total: _____ out of 42

Date: _____

Day of Week: _____

STEP 1 (1 min) Warm-up

Start the timer

Answer these.

$10 - 6 =$ ☐ $9 - 6 =$ ☐ $10 - 7 =$ ☐ $10 - 8 =$ ☐ $2 + 3 =$ ☐

$5 - 0 =$ ☐ $9 - 7 =$ ☐ $6 - 5 =$ ☐ $9 - 8 =$ ☐ $6 - 3 =$ ☐

STEP 2 (2.5 min) Rapid calculation

Start the timer

Answer these.

$6 - 1 =$ ☐ $5 + 2 =$ ☐ $1 + 0 =$ ☐ $9 - 5 =$ ☐ $6 + 1 =$ ☐

$2 + 5 =$ ☐ $2 + 6 =$ ☐ $2 + 2 =$ ☐ $5 - 2 =$ ☐ $10 - 1 =$ ☐

$3 + 3 =$ ☐ $6 - 2 =$ ☐ $5 - 3 =$ ☐ $9 - 2 =$ ☐ $2 + 7 =$ ☐

$10 - 2 =$ ☐ $8 - 3 =$ ☐ $7 - 5 =$ ☐ $2 + 1 =$ ☐ $9 + 1 =$ ☐

STEP 3 (1.5 min) Challenge

Start the timer

Fill in the missing numbers.

$7 +$ ☐ $= 9$ ☐ $- 6 = 2$ $6 -$ ☐ $= 2$ $10 -$ ☐ $= 7$

$5 +$ ☐ $= 6$ $1 +$ ☐ $= 4$ $3 -$ ☐ $= 0$ ☐ $+ 4 = 10$

☐ $+ 1 = 8$ ☐ $+ 6 = 9$ ☐ $- 0 = 8$ ☐ $+ 5 = 6$

Time spent: _____ min _____ sec. Total: _____ out of 42

Date: _____

Day of Week: _____

STEP 1 (1 min) **Warm-up**

Start the timer

Answer these.

3 + 4 = ☐   4 + 2 = ☐   7 + 3 = ☐   2 + 6 = ☐   5 + 2 = ☐

13 + 4 = ☐   14 + 2 = ☐   7 + 13 = ☐   2 + 16 = ☐   5 + 12 = ☐

STEP 2 (2.5 min) **Rapid calculation**

Start the timer

Answer these.

4 + 12 = ☐   5 + 13 = ☐   5 + 15 = ☐   13 + 5 = ☐

3 + 13 = ☐   6 + 12 = ☐   13 + 3 = ☐   3 + 12 = ☐

14 + 3 = ☐   1 + 13 = ☐   12 + 7 = ☐   3 + 17 = ☐

12 + 3 = ☐   1 + 14 = ☐   15 + 5 = ☐   2 + 11 = ☐

STEP 3 (1.5 min) **Challenge**

Start the timer

Fill in the missing numbers.

16 + ☐ = 18   13 + ☐ = 19   12 + ☐ = 17   ☐ + 12 = 20

14 + ☐ = 15   ☐ + 14 = 18   3 + ☐ = 14   ☐ + 15 = 16

5 + ☐ = 13   4 + ☐ = 11

Date: _____

Day of Week: _____

STEP 1 🕐 1 min **Warm-up**

Start the timer

🖱 **TIP** *Addition with carry: observe the larger number, break the smaller one down and then regroup with the larger number to whole tens.*

Answer these.

8 + 4 = ☐   9 + 6 = ☐   5 + 6 = ☐   5 + 8 = ☐   8 + 6 = ☐

6 + 7 = ☐   3 + 8 = ☐   4 + 9 = ☐   5 + 9 = ☐   9 + 7 = ☐

STEP 2 🕐 2.5 min **Rapid calculation**

Start the timer

Answer these.

8 + 7 = ☐   9 + 3 = ☐   8 + 3 = ☐   4 + 7 = ☐   8 + 5 = ☐

6 + 9 = ☐   6 + 6 = ☐   3 + 9 = ☐   8 + 4 = ☐   6 + 5 = ☐

7 + 9 = ☐   8 + 7 = ☐   9 + 2 = ☐   5 + 7 = ☐   7 + 5 = ☐

6 + 8 = ☐   7 + 7 = ☐   6 + 9 = ☐   7 + 8 = ☐   3 + 9 = ☐

STEP 3 🕐 1.5 min **Challenge**

Start the timer

Fill in the missing numbers.

9 + ☐ = 13   ☐ + 7 = 12   ☐ + 9 = 11   ☐ + 5 = 14

☐ + 4 = 12   8 + ☐ = 14   9 + ☐ = 17   ☐ + 6 = 13

8 + ☐ = 12   ☐ + 7 = 14

Time spent: _____ min _____ sec. Total: _____ out of 40

Date: _____

Day of Week: _____

STEP 1 (1 min) Warm-up

Start the timer

Answer these.

7 – 3 = ☐   8 – 6 = ☐   5 – 2 = ☐   9 – 5 = ☐   7 – 4 = ☐

17 – 3 = ☐   18 – 6 = ☐   15 – 2 = ☐   19 – 5 = ☐   17 – 4 = ☐

STEP 2 (2.5 min) Rapid calculation

Start the timer

Answer these.

16 – 4 = ☐   16 – 5 = ☐   18 – 3 = ☐   18 – 2 = ☐

16 – 3 = ☐   19 – 3 = ☐   17 – 6 = ☐   15 – 3 = ☐

12 – 6 = ☐   18 – 4 = ☐   19 – 5 = ☐   15 – 4 = ☐

19 – 8 = ☐   17 – 2 = ☐   18 – 7 = ☐   14 – 3 = ☐

STEP 3 (1.5 min) Challenge

Start the timer

Fill in the missing numbers.

19 – ☐ = 13   18 – ☐ = 13   17 – ☐ = 13   16 – ☐ = 13

17 – ☐ = 12   ☐ – 3 = 11   19 – ☐ = 11   ☐ – 2 = 11

☐ – 3 = 15   19 – ☐ = 14

Time spent: _____ min _____ sec. Total: _____ out of 36

STEP 1 (1 min) Warm-up

Start the timer

Answer these.

11 − 2 = ☐   12 − 3 = ☐   13 − 4 = ☐   14 − 5 = ☐

15 − 6 = ☐   12 − 9 = ☐   14 − 9 = ☐   18 − 9 = ☐

STEP 2 (2.5 min) Rapid calculation

Start the timer

Answer these.

15 − 9 = ☐   11 − 3 = ☐   12 − 5 = ☐   13 − 5 = ☐

15 − 7 = ☐   11 − 7 = ☐   13 − 8 = ☐   11 − 3 = ☐

13 − 6 = ☐   15 − 9 = ☐   14 − 6 = ☐   12 − 4 = ☐

14 − 7 = ☐   10 − 5 = ☐   12 − 6 = ☐   16 − 8 = ☐

13 − 7 = ☐   11 − 5 = ☐

STEP 3 (1.5 min) Challenge

Start the timer

Fill in the missing numbers.

15 − ☐ = 8   ☐ − 5 = 9   ☐ − 7 = 7   11 − ☐ = 9

12 − ☐ = 6   14 − ☐ = 7   16 − ☐ = 8   ☐ − 9 = 9

☐ − 6 = 9   12 − ☐ = 5

Time spent: _____ min _____ sec. Total: _____ out of 36

STEP 1 1 min **Warm-up**

Start the timer

Answer these.

2 + 8 = ☐   9 + 4 = ☐   6 + 7 = ☐   9 + 8 = ☐

10 – 3 = ☐   13 – 2 = ☐   13 – 1 = ☐   17 – 5 = ☐

STEP 2 2.5 min **Rapid calculation**

Start the timer

Answer these.

13 – 4 = ☐   9 + 5 = ☐   15 – 3 = ☐   9 + 6 = ☐

9 + 7 = ☐   18 – 5 = ☐   8 + 4 = ☐   17 – 4 = ☐

16 – 0 = ☐   9 + 4 = ☐   5 + 7 = ☐   16 – 5 = ☐

7 + 5 = ☐   11 – 4 = ☐   9 + 8 = ☐   12 – 6 = ☐

8 + 6 = ☐   14 – 2 = ☐

STEP 3 1.5 min **Challenge**

Start the timer

Fill in the missing numbers.

15 – ☐ = 12   7 + ☐ = 11   8 + ☐ = 10   12 – ☐ = 11

16 – ☐ = 10   ☐ – 3 = 12   ☐ + 3 = 16   8 + ☐ = 19

☐ + 5 = 14   6 + ☐ = 15

25 Mixed addition and subtraction for numbers up to 20 (2)

Date: _____

Day of Week: _____

Start the timer

Answer these.

5 + 8 = ☐   7 + 8 = ☐   5 + 7 = ☐   1 + 9 = ☐

13 − 3 = ☐   15 − 4 = ☐   12 − 2 = ☐   10 − 6 = ☐

STEP 2 (2.5 min) **Rapid calculation**

Start the timer

Answer these.

16 − 6 = ☐   11 − 0 = ☐   18 − 7 = ☐   13 − 2 = ☐

7 + 7 = ☐   9 + 9 = ☐   8 + 4 = ☐   5 + 9 = ☐

17 − 5 = ☐   9 + 5 = ☐   15 − 5 = ☐   6 + 8 = ☐

8 + 6 = ☐   15 − 3 = ☐   3 + 8 = ☐   18 − 7 = ☐

9 + 7 = ☐   16 − 5 = ☐

STEP 3 (1.5 min) **Challenge**

Start the timer

Fill in the missing numbers.

9 − ☐ = 3   16 − ☐ = 12   3 + ☐ = 14   ☐ + 5 = 12

13 − ☐ = 7   ☐ + 3 = 11   ☐ − 2 = 12   7 + ☐ = 14

☐ + 8 = 12   11 + ☐ = 19

Time spent: _____ min _____ sec. Total: _____ out of 36   ©HarperCollinsPublishers 2019

STEP 1 (1 min) **Warm-up**

Start the timer

Answer these.

$6 + 6 = \boxed{}$ $7 + 3 = \boxed{}$ $9 + 8 = \boxed{}$ $6 + 4 = \boxed{}$

$12 - 2 = \boxed{}$ $10 - 5 = \boxed{}$ $17 - 5 = \boxed{}$ $10 - 3 = \boxed{}$

STEP 2 (2.5 min) **Rapid calculation**

Start the timer

Answer these.

$12 - 3 = \boxed{}$ $6 + 5 = \boxed{}$ $15 - 3 = \boxed{}$ $16 + 4 = \boxed{}$

$10 - 7 = \boxed{}$ $4 + 8 = \boxed{}$ $16 - 7 = \boxed{}$ $2 + 9 = \boxed{}$

$15 - 6 = \boxed{}$ $5 + 10 = \boxed{}$ $10 - 8 = \boxed{}$ $15 + 3 = \boxed{}$

$11 - 4 = \boxed{}$ $8 + 6 = \boxed{}$ $11 - 2 = \boxed{}$ $5 + 8 = \boxed{}$

$5 + 9 = \boxed{}$ $14 - 3 = \boxed{}$

STEP 3 (1.5 min) **Challenge**

Start the timer

Fill in the missing numbers.

$13 - \boxed{} = 11$ $9 + \boxed{} = 11$ $12 - \boxed{} = 10$ $\boxed{} + 8 = 11$

$16 - \boxed{} = 8$ $\boxed{} + 9 = 15$ $15 - \boxed{} = 12$ $10 - \boxed{} = 3$

$\boxed{} + 8 = 16$ $9 + \boxed{} = 17$

Time spent: _____ min _____ sec. Total: _____ out of 36

Date: _____

Day of Week: _____

STEP 1 (1 min) Warm-up

Start the timer

Write the number of shapes for each of these.

⊚⊚⊚⊚⊚⊚⊚⊚⊚⊚ ⊚ ☐   ▲▲▲▲▲▲▲▲▲▲ ▲▲▲▲ ☐   ◇◇◇◇◇◇◇◇◇◇ ◇◇◇◇◇ ☐

★★★★★★★★★★ ★★★★★★★★★ ☐   ■■■■■■■■■■ ■■■■■■■■■■ ☐   ○○○○○○○○○○ ○○ ☐

STEP 2 (2.5 min) Rapid calculation

Start the timer

Complete each number sentence with '**is less than**', '**is greater than**', or '**equals**'.

1. 13 _____ 14   **2.** 17 _____ 16

3. 18 _____ 19   **4.** 12 _____ 6 + 6

5. 15 _____ 9 + 7   **6.** 8 + 5 _____ 12

Arrange each set of numbers from **smallest** to **largest** (<).

7. 11, 10, 15, 17, 19

☐ < ☐ < ☐ < ☐ < ☐

8. 20, 15, 18, 14, 13

☐ < ☐ < ☐ < ☐ < ☐

Arrange each set of numbers from **largest** to **smallest** (>).

9. 14, 18, 10, 17, 13

☐ > ☐ > ☐ > ☐ > ☐

10. 15, 20, 14, 19, 12

☐ > ☐ > ☐ > ☐ > ☐

STEP 3 (1.5 min) Challenge

Start the timer

Write a number to complete each statement.

1. 17 is greater than ☐   **2.** 16 is less than ☐   **3.** 18 is less than ☐

4. ☐ is greater than 16   **5.** ☐ is less than 19

6. 14 is one more than ☐   **7.** 13 is one less than ☐

8. ☐ is one less than 20   **9.** ☐ is one more than 11

Time spent: _____ min _____ sec. Total: _____ out of 25

STEP 1 (1 min) **Warm-up** Start the timer

Write the number represented by the counters in each place value chart.

Tens	Ones
●	●

☐

Tens	Ones
●	●●●●●

☐

Tens	Ones
●	●●●●● ●

☐

Tens	Ones
●	●●●

☐

Tens	Ones
●	●●

☐

Tens	Ones
●	●●●●● ●●

☐

Tens	Ones
●	●●●●● ●●●●

☐

Tens	Ones
●●	

☐

STEP 2 (2.5 min) **Rapid calculation** Start the timer

Answer these.

$10 + 1 = $ ☐ $10 + 3 = $ ☐ $10 + 2 = $ ☐ $10 + 6 = $ ☐

$10 + 4 = $ ☐ $10 + 9 = $ ☐ $10 + 7 = $ ☐ $10 + 10 = $ ☐

$11 - 1 = $ ☐ $13 - 3 = $ ☐ $17 - 7 = $ ☐ $18 - 8 = $ ☐

$16 - 6 = $ ☐ $12 - 2 = $ ☐ $14 - 4 = $ ☐ $19 - 9 = $ ☐

STEP 3 (1.5 min) **Challenge** Start the timer

Fill in the missing numbers.

$17 = 10 + $ ☐ $16 = 10 + $ ☐ $14 = 10 + $ ☐

$10 + $ ☐ $= 16$ $10 + $ ☐ $= 13$ $10 + $ ☐ $= 20$

$15 - $ ☐ $= 10$ $20 - $ ☐ $= 10$ ☐ $- 2 = 10$

Date: _____

Day of Week: _____

STEP 1 (1 min) Warm-up

Start the timer

Fill in the missing numbers in each number line.

| 11 | 12 | | |

| | 15 | 16 | |

| 13 | | | 17 |

| | 17 | | 20 |

| 14 | | 12 | |

| 17 | 16 | | |

| | | 16 | 15 |

| 20 | | | 16 |

STEP 2 (2.5 min) Rapid calculation

Start the timer

Fill in the missing numbers in each number line.

| 7 | | | 10 | | |

| 20 | | | 16 | |

| | | | 15 | | 18 |

| | 17 | | | 14 |

| 18 | | | | 12 |

| 9 | 10 | | | | 15 |

| | | 16 | | 13 | | |

| | | 19 | 17 | | | |

| 9 | | | | | | 18 |

| | | | 14 | | 11 | | |

STEP 3 (1.5 min) Challenge

Start the timer

Fill in the missing numbers in each sequence.

13, ☐, 17, 19, ☐, ☐

20, ☐, ☐, 11, ☐, 5

7, ☐, 15, ☐, ☐, 27

10, ☐, ☐, 16, ☐, ☐

☐, 18, ☐, ☐, 9, ☐

25, ☐, 15, ☐, 5, ☐

☐, ☐, ☐, 12, 8, 4

☐, ☐, ☐, 13, 8, 3

☐, ☐, 17, ☐, 7, ☐

Time spent: _____ min _____ sec. Total: _____ out of 27

STEP 1 (1 min) **Warm-up**

Start the timer

Answer these.

7 – 3 = ☐   8 – 6 = ☐   5 – 2 = ☐   9 – 5 = ☐

7 – 4 = ☐   10 – 2 = ☐   17 – 10 = ☐   20 – 10 = ☐

STEP 2 (2.5 min) **Rapid calculation**

Start the timer

Answer these.

16 – 4 = ☐   15 – 2 = ☐   18 – 3 = ☐   18 – 7 = ☐

19 – 3 = ☐   16 – 5 = ☐   17 – 3 = ☐   17 – 6 = ☐

15 – 3 = ☐   17 – 7 = ☐   18 – 4 = ☐   19 – 5 = ☐

19 – 8 = ☐   18 – 5 = ☐   16 – 2 = ☐   14 – 3 = ☐

16 – 6 = ☐   18 – 8 = ☐   15 – 5 = ☐   14 – 2 = ☐

STEP 3 (1.5 min) **Challenge**

Start the timer

Complete each number sentence with 'is less than', 'is greater than', or 'equals'.

16 – 3 _____ 14 – 1   15 – 4 _____ 17 – 4

18 – 2 _____ 19 – 3   17 – 5 _____ 16 – 5

19 – 7 _____ 13 – 1   20 – 10 _____ 16 – 5

19 – 2 _____ 18 – 4   17 – 2 _____ 18 – 2

17 – 2 _____ 19 – 5

Time spent: _____ min _____ sec. Total: _____ out of 37

STEP 1 (1 min) Warm-up

Start the timer

Fill in the missing numbers. The first one has been done for you.

$7 + 5 =$ **12**
3 **2**
10

$8 + 6 = \boxed{}$

$3 + 9 = \boxed{}$

$9 + 5 = \boxed{}$

$4 + 7 = \boxed{}$

$6 + 9 = \boxed{}$

STEP 2 (2.5 min) Rapid calculation

Start the timer

Answer these.

$5 + 9 = \boxed{}$ $8 + 9 = \boxed{}$ $6 + 7 = \boxed{}$ $2 + 9 = \boxed{}$

$2 + 8 = \boxed{}$ $9 + 3 = \boxed{}$ $7 + 8 = \boxed{}$ $4 + 8 = \boxed{}$

$6 + 5 = \boxed{}$ $9 + 4 = \boxed{}$ $4 + 7 = \boxed{}$ $7 + 5 = \boxed{}$

$9 + 6 = \boxed{}$ $8 + 8 = \boxed{}$ $7 + 6 = \boxed{}$ $6 + 8 = \boxed{}$

STEP 3 (1.5 min) Challenge

Start the timer

Complete each number sentence with 'is less than', 'is greater than', or 'equals'.

$7 + 5$ _____ 12   11 _____ $4 + 9$

14 _____ $7 + 8$ $5 + 7$ _____ 13

$9 + 8$ _____ 16   15 _____ $9 + 7$

17 _____ $8 + 8$ $9 + 4$ _____ 13

16 _____ $9 + 6$

Time spent: _____ min _____ sec. Total: _____ out of 30

STEP 1 (1 min) **Warm-up**

Start the timer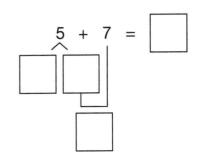

Fill in the missing numbers.

8 + 3 = ☐

9 + 4 = ☐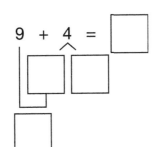

5 + 7 = ☐

9 + 8 = ☐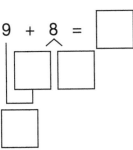

6 + 8 = ☐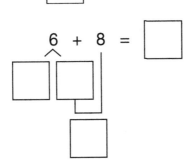

9 + 9 = ☐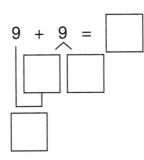

STEP 2 (2.5 min) **Rapid calculation**

Start the timer

Answer these.

8 + 3 = ☐   8 + 4 = ☐   8 + 5 = ☐   8 + 6 = ☐

8 + 7 = ☐   8 + 8 = ☐   8 + 9 = ☐   8 + 10 = ☐

9 + 2 = ☐   9 + 3 = ☐   9 + 4 = ☐   9 + 5 = ☐

9 + 6 = ☐   9 + 7 = ☐   9 + 8 = ☐   9 + 9 = ☐

STEP 3 (1.5 min) **Challenge**

Start the timer

Complete each number sentence with '**is less than**', '**is greater than**', or '**equals**'.

4 + 8 _____ 8 + 4   3 + 9 _____ 4 + 9

5 + 8 _____ 6 + 8   7 + 7 _____ 6 + 8

6 + 9 _____ 8 + 5   7 + 9 _____ 10 + 6

5 + 9 _____ 8 + 4   7 + 6 _____ 9 + 4

8 + 8 _____ 13 + 5

Time spent: _____ min _____ sec. Total: _____ out of 31

Date: _____

Day of Week: _____

STEP 1 (1 min) Warm-up

Fill in the missing numbers. The first one has been done for you.

14 − 5 = [9]

[4] [1]

[10]

12 − 9 = [ ]

[ ] [ ]

[ ]

13 − 6 = [ ]

[ ] [ ]

[ ]

16 − 8 = [ ]

[ ] [ ]

[ ]

15 − 7 = [ ]

[ ] [ ]

[ ]

11 − 3 = [ ]

[ ] [ ]

[ ]

STEP 2 (2.5 min) Rapid calculation

Answer these.

17 − 8 = [ ]   13 − 7 = [ ]   11 − 3 = [ ]   11 − 9 = [ ]

15 − 7 = [ ]   14 − 8 = [ ]   14 − 9 = [ ]   11 − 7 = [ ]

14 − 5 = [ ]   18 − 9 = [ ]   13 − 8 = [ ]   16 − 8 = [ ]

13 − 6 = [ ]   11 − 5 = [ ]   12 − 6 = [ ]   17 − 9 = [ ]

STEP 3 (1.5 min) Challenge

Complete each number sentence with 'is less than', 'is greater than', or 'equals'.

12 − 7 _____ 4   6 _____ 14 − 9

4 _____ 13 − 7   14 − 8 _____ 7

8 _____ 16 − 8   15 − 6 _____ 10

3 _____ 12 − 9   17 − 8 _____ 8

14 − 5 _____ 9

Time spent: _____ min _____ sec. Total: _____ out of 30   ©HarperCollinsPublishers 2019

STEP 1 (1 min) **Warm-up**

Start the timer

Fill in the missing numbers.

10 +3 / −3 □

15 +3 / −3 □

16 +4 / −4 □

17 −2 / +2 □

□ +5 / −5 17

□ +4 / −4 16

□ +3 / −□ 12

□ +□ / −6 18

STEP 2 (2.5 min) **Rapid calculation**

Start the timer

Answer these.

$8 + 6 =$ □

$9 + 4 =$ □

$5 + 7 =$ □

$8 + 9 =$ □

$14 - 6 =$ □

$13 - 4 =$ □

$12 - 7 =$ □

$17 - 9 =$ □

$6 + 7 =$ □

$4 + 8 =$ □

$9 + 6 =$ □

$14 + 6 =$ □

$13 - 7 =$ □

$12 - 8 =$ □

$15 - 6 =$ □

$20 - 6 =$ □

STEP 3 (1.5 min) **Challenge**

Start the timer

Answer the addition sums and then complete the inverse calculations.

1. $6 + 8 =$ □ $14 - $ □ $=$ □

2. $9 + 4 =$ □ $13 - $ □ $=$ □

3. $7 + 6 =$ □   □ $- 6 =$ □

Fill in the missing numbers.

4. □ $- 3 = 10 + 5$

5. □ $+ 6 = 20 - 4$

6. $18 - $ □ $= 6 + 9$

7. $6 + $ □ $= 17 - 8$

8. $19 - $ □ $= 8 - 3$

9. $13 - 13 = $ □ $- $ □

©HarperCollinsPublishers 2019

Time spent: _____ min _____ sec. Total: _____ out of 33

35 Number walls

Date: _____

Day of Week: _____

STEP 1 (1 min) Warm-up

Start the timer

Fill in the missing numbers.

9 + 2 = ☐   17 − 2 = ☐   6 + 4 = ☐   18 − 8 = ☐   7 + 5 = ☐

11 − 3 = ☐   6 + ☐ = 10   ☐ − 7 = 4   8 − ☐ = 1   ☐ − 2 = 16

STEP 2 (2.5 min) Rapid calculation

Start the timer

Complete the number walls. Each box is filled by adding the two numbers below it.

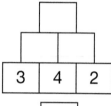 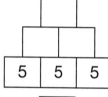 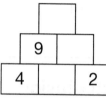 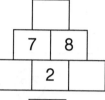

  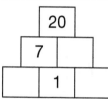 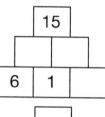 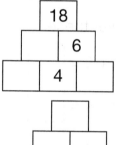

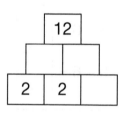 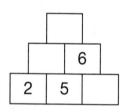 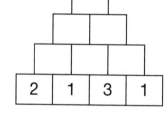 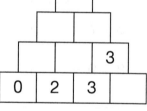

STEP 3 (1.5 min) Challenge

Start the timer

Complete these number walls.

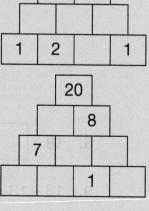 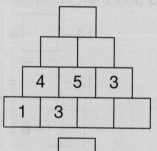 

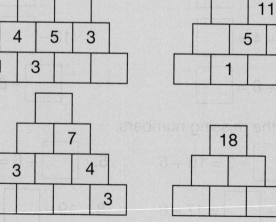

Time spent: _____ min _____ sec. Total: _____ out of 28

Date: _____

Day of Week: _____

Start the timer

STEP 1 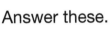 **Warm-up**

Answer these.

2 + 2 = ☐

2 + 3 = ☐

2 + 4 = ☐

2 + 5 = ☐

9 − 4 = ☐

9 − 5 = ☐

9 − 6 = ☐

9 − 7 = ☐

Start the timer

STEP 2 (2.5 min) **Rapid calculation**

Answer these.

2 + 6 = ☐

2 + 7 = ☐

9 − 2 = ☐

9 − 3 = ☐

10 + 3 = ☐

11 + 3 = ☐

12 + 3 = ☐

13 + 3 = ☐

6 + 6 = ☐

6 + 7 = ☐

6 + 8 = ☐

6 + 9 = ☐

14 − 4 = ☐

15 − 4 = ☐

16 − 4 = ☐

17 − 4 = ☐

16 − 6 = ☐

16 − 7 = ☐

16 − 8 = ☐

16 − 9 = ☐

Start the timer

STEP 3 (1.5 min) **Challenge**

Fill in the missing numbers.

8 + ☐ = 15

7 + ☐ = 15

6 + ☐ = 15

☐ + 7 = 10

☐ + 7 = 12

☐ + 7 = 14

14 − ☐ = 10

14 − ☐ = 8

14 − ☐ = 6

☐ − 10 = 10

☐ − 5 = 10

☐ − 7 = 10

©HarperCollinsPublishers 2019

Time spent: _____ min _____ sec. Total: _____ out of 40

39

37 Double and half

Date: _____

Day of Week: _____

STEP 1 (1 min) Warm-up

Start the timer

Answer these.

1 + 1 = ☐   2 + 2 = ☐   3 + 3 = ☐   4 + 4 = ☐

5 + 5 = ☐   6 + 6 = ☐   7 + 7 = ☐   8 + 8 = ☐

9 + 9 = ☐   10 + 10 = ☐   8 − 4 = ☐   20 − 10 = ☐

STEP 2 (2.5 min) Rapid calculation

Start the timer

Fill in the missing numbers.

2 Double→ ☐   4 Double→ ☐   5 Double→ ☐   8 Double→ ☐

10 Double→ ☐   9 Double→ ☐   7 Double→ ☐   3 Double→ ☐ Double→ ☐

6 Half→ ☐   12 Half→ ☐   8 Half→ ☐   18 Half→ ☐

4 Half→ ☐   14 Half→ ☐   16 Half→ ☐   20 Half→ ☐ Half→ ☐

STEP 3 (1.5 min) Challenge

Start the timer

Fill in the missing numbers.

☐ Double→ 8   ☐ Double→ 18   ☐ Double→ 10

☐ Double→ 12   ☐ Half→ 10   ☐ Half→ 6

☐ Half→ 2   ☐ Half→ 4   ☐ Double→ 4

Time spent: _____ min _____ sec. Total: _____ out of 39

Date: _____

Day of Week: _____

STEP 1 (1 min) **Warm-up**

Start the timer

Answer these.

5 + 2 = ☐   10 + 7 = ☐   15 + 5 = ☐   10 + 4 = ☐

6 + 3 = ☐   12 + 4 = ☐   3 + 7 = ☐   8 + 3 = ☐

STEP 2 (2.5 min) **Rapid calculation**

Start the timer

Answer these.

2 + 8 = ☐   13 + 4 = ☐   4 + 11 = ☐   9 + 6 = ☐

2 + 15 = ☐   1 + 12 = ☐   10 + 5 = ☐   7 + 7 = ☐

5 + 8 = ☐   11 + 8 = ☐   3 + 13 = ☐   14 + 4 = ☐

2 + 14 = ☐   12 + 5 = ☐   1 + 16 = ☐   4 + 15 = ☐

11 + 9 = ☐   13 + 6 = ☐   6 + 13 = ☐   18 + 2 = ☐

STEP 3 (1.5 min) **Challenge**

Start the timer

Fill in the missing numbers.

10 + ☐ = 12   ☐ + 10 = 16   ☐ + 10 = 19

10 + ☐ = 14   ☐ + 10 = 18   10 + ☐ = 20

☐ + 7 = 17   11 + ☐ = 16   10 + ☐ = 18

Time spent: _____ min _____ sec. Total: _____ out of 37

39 Subtracting numbers up to 20

Date: _____

Day of Week: _____

STEP 1 (1 min) Warm-up

Start the timer

Answer these.

6 − 4 = ☐   9 − 2 = ☐   10 − 7 = ☐   17 − 7 = ☐

16 − 6 = ☐   15 − 5 = ☐   18 − 8 = ☐   11 − 1 = ☐

STEP 2 (2.5 min) Rapid calculation

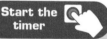

Start the timer

Answer these.

13 − 3 = ☐   12 − 2 = ☐   14 − 4 = ☐   20 − 10 = ☐

14 − 12 = ☐   13 − 9 = ☐   12 − 9 = ☐   15 − 9 = ☐

16 − 14 = ☐   11 − 4 = ☐   18 − 17 = ☐   17 − 15 = ☐

12 − 10 = ☐   14 − 10 = ☐   15 − 10 = ☐   13 − 10 = ☐

19 − 9 = ☐   18 − 14 = ☐   16 − 13 = ☐   10 − 0 = ☐

STEP 3 (1.5 min) Challenge

Start the timer

Fill in the missing numbers.

☐ − 10 = 8   ☐ − 10 = 2   ☐ − 11 = 7

10 − ☐ = 6   10 − ☐ = 8   10 − ☐ = 5

10 − ☐ = 1   ☐ − 10 = 6   5 − ☐ = 4

Time spent: _____ min _____ sec. Total: _____ out of 37   ©HarperCollinsPublishers 2019

STEP 1 (1 min) Warm-up

Start the timer

Answer these.

$2 + 8 = \boxed{}$ $9 + 4 = \boxed{}$ $6 + 7 = \boxed{}$ $9 + 8 = \boxed{}$

$8 + 6 = \boxed{}$ $10 - 3 = \boxed{}$ $13 - 2 = \boxed{}$ $13 - 1 = \boxed{}$

STEP 2 (2.5 min) Rapid calculation

Start the timer

Answer these.

$14 - 4 = \boxed{}$ $12 + 2 = \boxed{}$ $15 - 3 = \boxed{}$ $7 + 5 = \boxed{}$

$10 - 6 = \boxed{}$ $9 + 4 = \boxed{}$ $14 - 2 = \boxed{}$ $8 + 5 = \boxed{}$

$11 - 7 = \boxed{}$ $3 + 8 = \boxed{}$ $13 - 3 = \boxed{}$ $5 + 4 = \boxed{}$

$12 - 2 = \boxed{}$ $7 + 3 = \boxed{}$ $11 - 1 = \boxed{}$ $10 + 6 = \boxed{}$

$17 - 5 = \boxed{}$ $19 - 7 = \boxed{}$ $2 + 8 - 4 = \boxed{}$ $9 + 7 - 6 = \boxed{}$

STEP 3 (1.5 min) Challenge

Start the timer

Complete each number sentence with 'is less than', 'is greater than', or 'equals'.

$15 - 6$ _____ 10 $5 + 3$ _____ 8

9 _____ $17 - 7$ $12 - 7$ _____ 4

$14 - 7$ _____ 6 7 _____ $6 + 3$

5 _____ $11 - 4$ 18 _____ $4 + 16$

6 _____ $13 - 7$

Answers are given from top left, left to right, unless otherwise stated.

Test 1

Step 1:

1. 3	**2.** 2	**3.** 4
4. 1	**5.** 2	**6.** 5
7. 4	**8.** 3	**9.** 5

Step 2:

1. △△△△△ **2.** △△△
3. △△△△ **4.** △△
5. 3 pineapples circled and 3 strawberries circled
6. 4 cherries circled and 2 apples circled
7. 3 bananas circled and 5 lemons circled
8. 2 oranges circled and 5 bananas circled
9. 1 pineapple circled and 3 strawberries circled
10. 4 oranges circled and 4 apples circled

Step 3:

4; 2; 3; 2; (2 + 2 =) 4

Test 2

Step 1:

1. 6	**2.** 8	**3.** 7
4. 9	**5.** 10	**6.** 9
7. 7	**8.** 6	**9.** 8

Step 2:

1. △△△△△△△ **2.** △△△△△
3. △△△△△△△△ **4.** △△△△△△△△△△
5. 8 pineapples circled and 7 strawberries circled
6. 9 cherries circled and 10 apples circled
7. 8 bananas circled and 6 lemons circled
8. 5 oranges circled and 10 bananas circled
9. 6 pineapples circled and 9 strawberries circled
10. 10 oranges circled and 6 apples circled

Step 3:

8; 10; 9; 5; 9

Test 3

Step 1:

 – 1 – One 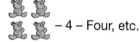 – 4 – Four, etc.

Step 2:

1. 4 dark fish circled and 3 pale fish circled
2. 3 cherries circled and 2 apples circled
3. 3 bananas circled and 5 lemons circled

4.

5.

6.

Step 3:

shape; colour; size (or any other suitable answer)

Test 4

Step 1:

 – 1 – One  – 4 – Four, etc.

Step 2:

●●○○○; Two; 2 ●●●●●
●●○○○; Seven; 7, etc.

Step 3:

○△; ○○△; ○○○○△

Test 5

Step 1:

8 coloured in; 6 coloured in;
9 coloured in; 7 coloured in; 10 coloured in

Step 2:

1. ○

2.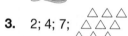

3. 2; 4; 7; △△△ △△△ △△△

Step 3:

Ruler matched to third box; gloves matched to second box; football matched to first box; doll matched to fifth box; small ball matched to fourth box

Test 6

Step 1:

2; 4; 5; 6; 7; 8; 9

Step 2:

1. 4; 5; 7; 6th; 1st; 5th
2. 10; 5; 8; 5th; 3rd; 7th

Step 3:

Test 7

Step 1:

2; 5; 6; 9, 7, 6, 4, 3, 2, 1; 7; 5; 8; 1, 2, 4, 5, 6, 7, 8, 10

Answers

Step 2:
1. Any answer 5 or below
2. Any answer 7 or below
3. Any answer 3 or above
4. Any answer 6 or above
5. Any answer 8 or below
6. Any answer 2 or above
7. 2, 3, 5, 8, 9
8. 3, 5, 7, 9, 10
9. 0, 3, 5, 7, 9
10. 10, 8, 7, 4, 1
11. 9, 5, 4, 2, 0
12. 8, 6, 5, 3, 1

Step 3:
1. Any answers 6 or below
2. Any answers 8 or above
3. Any answers 4 or below
4. Any answers 6 or above
5. Any answers 2 or below
6. Any answers 4 or above

Test 8
Step 1:
1, 4; 3, 2; 4, 1; 3, 3; 2, 4; 5, 1; 4, 2; 1, 5

Step 2:
1st col: 1, 4; 3, 2; 2, 3; 3, 2; 2, 3; 3, 2; 1, 4; 2, 3; 1, 4; 2, 3
2nd col: 3, 3; 2, 4; 4, 2; 3, 3; 2, 4; 4, 2; 4, 2; 3, 3; 4, 2; 2, 4

Step 3:
4; 2; 1; 3; 5; 4; 2; 3; 1; 5

Test 9
Step 1:
3, 4; 2, 5; 4, 3; 6, 1; 5, 2; 5, 3; 3, 5; 7, 1; 4, 4; 2, 6

Step 2:
1st col: 2, 5; 3, 4; 4, 3; 1, 6; 3, 4; 5, 2; 6, 1; 4, 3; 1, 6; 2, 5
2nd col: 3, 5; 2, 6; 6, 2; 4, 4; 1, 7; 7, 1; 5, 3; 4, 4; 5, 3; 2, 6

Step 3:
1; 5; 2; 4; 6; 7; 1; 5; 4; 6

Test 10
Step 1:
3, 6; 2, 7; 1, 8; 5, 4; 7, 2; 7, 3; 5, 5; 9, 1; 8, 2; 4, 6

Step 2:
1st col: 5, 4; 3, 6; 1, 8; 2, 7; 4, 5; 7, 2; 6, 3; 8, 1; 5, 4; 7, 2
2nd col: 5, 5; 4, 6; 8, 2; 3, 7; 1, 9; 4, 6; 1, 9; 5, 5; 8, 2; 3, 7

Step 3:
5; 4; 6; 3; 8; 5; 9; 3; 8; 4

Test 11
Step 1:
2; 5; 5; 4; 4; 3; 3; 5; 4; 5

Step 2:
4; 2; 5; 1; 4; 5; 4; 3; 5; 5; 3; 5; 2; 2; 3; 4; 4; 3; 5; 5

Step 3:
2 + 1 = 3; 4 + 1 = 5; 1 + 4 = 5; 3 + 1 = 4;
3 + 2 = 5; 1 + 3 = 4; 2 + 3 = 5; 1 + 2 = 3

Test 12
Step 1:
6; 3; 6; 9; 8; 6; 4; 7; 3; 7

Step 2:
10; 8; 7; 10; 1; 9; 4; 8; 10; 9; 7; 7; 8; 7; 5; 9; 5; 10; 3; 9

Step 3:
3 + 2 = 5; 6 + 2 = 8; 4 + 4 = 8; 4 + 6 = 10;
4 + 3 = 7; 2 + 7 = 9; 2 + 1 = 3; 1 + 8 = 9; 1 + 5 = 6

Test 13
Step 1:
10; 8; 5; 10; 8; 5; 9; 9; 9; 10

Step 2:
8; 8; 10; 7; 9; 4; 6; 2; 4; 6; 4; 9; 9; 10; 7; 8; 3; 3; 7; 7

Step 3:
3 + 6 = 9; 7 + 2 = 9; 2 + 3 = 5; 4 + 2 = 6; 4 + 4 = 8;
2 + 8 = 10; 6 + 4 = 10; 6 + 3 = 9

Test 14
Step 1:
8; 6; 10; 7; 6; 7; 8; 9; 5; 4

Step 2:
5; 9; 3; 8; 9; 2; 6; 9; 3; 7; 6; 6; 10; 10; 9; 4; 5; 7; 8; 10

Step 3:
4 + 2 = 6; 2 + 7 = 9; 2 + 8 = 10; 3 + 5 = 8;
9 + 1 = 10; 6 + 4 = 10; 2 + 3 = 5; 1 + 8 = 9

Test 15
Step 1:
1; 1; 3; 3; 1; 2; 1; 2; 4; 2

Step 2:
3; 3; 2; 0; 2; 4; 0; 1; 1; 3; 2; 1; 1; 4; 0; 0; 1; 0; 2; 0

Step 3:
3; 1; 3; 4; 4; 5; 1; 4; 4

Test 16
Step 1:
3; 6; 3; 5; 3; 2; 1; 2; 4; 8

Step 2:
1; 1; 6; 6; 1; 2; 4; 10; 2; 8; 6; 2; 2; 5; 7; 5; 7; 3; 4; 3

Step 3:
3; 6; 5; 8; 3; 10; 2; 3

Mind Gym:
Any suitable answer, e.g.

Answers

Test 17

Step 1:

6; 5; 4; 2; 2; 2; 1; 1; 3; 8

Step 2:

2; 7; 8; 3; 2; 3; 6; 1; 1; 8; 3; 2; 2; 1; 4; 4; 5; 1; 10; 4

Step 3:

10; 4; 7; 10; 1; 2; 3; 4

Mind Gym:

Any suitable answer, e.g.

Test 18

Step 1:

3; 4; 8; 2; 1; 4; 5; 6; 9; 7

Step 2:

1; 2; 3; 3; 5; 1; 3; 2; 1; 1; 4; 3; 1; 6; 4; 1; 5; 8; 2; 5

Step 3:

2; 1; 7; 5; 0; 0; 4; 2; 6; 3; 8; 9

Test 19

Step 1:

4; 3; 3; 2; 5; 5; 2; 1; 1; 3

Step 2:

5; 7; 1; 4; 7; 7; 8; 4; 3; 9; 6; 4; 2; 7; 9; 8; 5; 2; 3; 10

Step 3:

2; 8; 4; 3; 1; 3; 3; 6; 7; 3; 8; 1

Test 20

Step 1:

7; 6; 10; 8; 7; 17; 16; 20; 18; 17

Step 2:

16; 18; 20; 18; 16; 18; 16; 15; 17; 14; 19; 20; 15; 15; 20; 13

Step 3:

2; 6; 5; 8; 1; 4; 11; 1; 8; 7

Test 21

Step 1:

12; 15; 11; 13; 14; 13; 11; 13; 14; 16

Step 2:

15; 12; 11; 11; 13; 15; 12; 12; 12; 11; 16; 15; 11; 12; 12; 14; 14; 15; 15; 12

Step 3:

4; 5; 2; 9; 8; 6; 8; 7; 4; 7

Test 22

Step 1:

4; 2; 3; 4; 3; 14; 12; 13; 14; 13

Step 2:

12; 11; 15; 16; 13; 16; 11; 12; 6; 14; 14; 11; 11; 15; 11; 11

Step 3:

6; 5; 4; 3; 5; 14; 8; 13; 18; 5

Test 23

Step 1:

9; 9; 9; 9; 9; 3; 5; 9

Step 2:

6; 8; 7; 8; 8; 4; 5; 8; 7; 6; 8; 8; 7; 5; 6; 8; 6; 6

Step 3:

7; 14; 14; 2; 6; 7; 8; 18; 15; 7

Test 24

Step 1:

10; 13; 13; 17; 7; 11; 12; 12

Step 2:

9; 14; 12; 15; 16; 13; 12; 13; 16; 13; 12; 11; 12; 7; 17; 6; 14; 12

Step 3:

3; 4; 2; 1; 6; 15; 13; 11; 9; 9

Test 25

Step 1:

13; 15; 12; 10; 10; 11; 10; 4

Step 2:

10; 11; 11; 11; 14; 18; 12; 14; 12; 14; 10; 14; 14; 12; 11; 11; 16; 11

Step 3:

6; 4; 11; 7; 6; 8; 14; 7; 4; 8

Test 26

Step 1:

12; 10; 17; 10; 10; 5; 12; 7

Step 2:

9; 11; 12; 20; 3; 12; 9; 11; 9; 15; 2; 18; 7; 14; 9; 13; 14; 11

Step 3:

2; 2; 2; 3; 8; 6; 3; 7; 8; 8

Test 27

Step 1:

11; 14; 15; 19; 20; 12

Step 2:

1. is less than   **2.** is greater than

3. is less than   **4.** equals

5. is less than   **6.** is greater than

7. 10 < 11 < 15 < 17 < 19

8. 13 < 14 < 15 < 18 < 20

9. 18 > 17 > 14 > 13 > 10

10. 20 > 19 > 15 > 14 > 12

Answers

Step 3:
1. Any answer 16 or below
2. Any answer 17 or above
3. Any answer 19 or above
4. Any answer 17 or above
5. Any answer 18 or below
6. 13   7. 14   8. 19   9. 12

Test 28
Step 1:
11; 15; 16; 13; 12; 17; 19; 20
Step 2:
11; 13; 12; 16; 14; 19; 17; 20; 10; 10; 10; 10; 10; 10; 10; 10
Step 3:
7; 6; 4; 6; 3; 10; 5; 10; 12

Test 29
Step 1:
13, 14, 15; 14, 17, 18; 14, 15, 16; 16, 18, 19; 13, 11, 10; 15, 14, 13; 19, 18, 17; 19, 18, 17
Step 2:
8, 9, 11, 12, 13; 19, 18, 17, 15, 14; 12, 13, 14, 16, 17; 19, 18, 16, 15, 13; 17, 16, 15, 14, 13; 11, 12, 13, 14; 19, 18, 17, 15, 14, 12, 11, 10; 20, 18, 16, 15, 14, 13, 12, 11; 10, 11, 12, 13, 14, 15, 16, 17; 17, 16, 15, 13, 12, 10, 9, 8
Step 3:
15, 21, 23; 20, 10, 0; 17, 14, 8; 24, 20, 16; 11, 19, 23; 28, 23, 18; 12, 14, 18, 20; 27, 22, 12, 2; 21, 15, 12, 6

Test 30
Step 1:
4; 2; 3; 4; 3; 8; 7; 10
Step 2:
12; 13; 15; 11; 16; 11; 14; 11; 12; 10; 14; 14; 11; 13; 14; 11; 10; 10; 10; 12
Step 3:
equals; is less than; equals; is greater than; equals; is less than; is greater than; is less than; is greater than

Test 31
Step 1:
(final result given last): 10, 2, 4, 14; 2, 1, 10, 12; 10, 1, 4, 14; 1, 3, 10, 11; 5, 1, 10, 15
Step 2:
14; 17; 13; 11; 10; 12; 15; 12; 11; 13; 11; 12; 15; 16; 13; 14
Step 3:
equals; is less than; is less than; is less than; is greater than; is less than; is greater than; equals; is greater than

Test 32
Step 1:
(final result given last): 10, 2, 1, 11; 10, 1, 3, 13; 2, 3, 10, 12; 10, 1, 7, 17; 4, 2, 10, 14; 10, 1, 8, 18
Step 2:
11; 12; 13; 14; 15; 16; 17; 18; 11; 12; 13; 14; 15; 16; 17; 18
Step 3:
equals; is less than; is less than; equals; is greater than; equals; is greater than; equals; is less than

Test 33
Step 1:
(final result given last): 10, 2, 7, 3; 10, 3, 3, 7; 10, 6, 2, 8; 10, 5, 2, 8; 10, 1, 2, 8
Step 2:
9; 6; 8; 2; 8; 6; 5; 4; 9; 9; 5; 8; 7; 6; 6; 8
Step 3:
is greater than; is greater than; is less than; is less than; equals; is less than; equals; is greater than; equals

Test 34
Step 1:
13; 18; 20; 15; 12; 12; 9, 3; 12, 6
Step 2:
14; 8; 13; 6; 13; 9; 12; 4; 12; 5; 15; 9; 17; 8; 20; 14
Step 3:
1. 14; 8, 6   2. 13; 4, 9
3. 13; 13, 7   4. 18
5. 10   6. 3
7. 3   8. 14
9. Any suitable answer, e.g. 20, 20

Test 35
Step 1:
11; 15; 10; 10; 12; 8; 4; 11; 7; 18
Step 2:

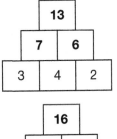
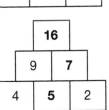
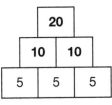
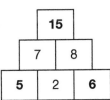

Answers

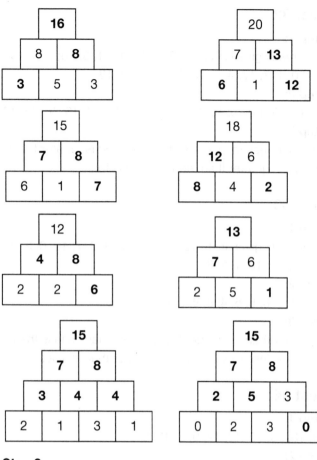

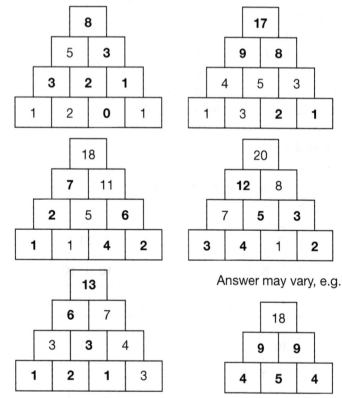

Step 3:

Answer may vary, e.g.

Test 36

Step 1:
4; 5; 6; 7; 5; 4; 3; 2

Step 2:
8; 9; 7; 6; 13; 14; 15; 16; 12; 13; 14; 15; 10; 11; 12; 13; 10; 9; 8; 7

Step 3:
7; 8; 9; 3; 5; 7; 4; 6; 8; 20; 15; 17

Test 37

Step 1:
2; 4; 6; 8; 10; 12; 14; 16; 18; 20; 4; 10

Step 2:
4; 8; 10; 16; 20; 18; 14; 6; 12; 3; 6; 4; 9; 2; 7; 8; 10, 5

Step 3:
4; 9; 5; 6; 20; 12; 4; 8; 2

Test 38

Step 1:
7; 17; 20; 14; 9; 16; 10; 11

Step 2:
10; 17; 15; 15; 17; 13; 15; 14; 13; 19; 16; 18; 16; 17; 17; 19; 20; 19; 19; 20

Step 3:
2; 6; 9; 4; 8; 10; 10; 5; 8

Test 39

Step 1:
2; 7; 3; 10; 10; 10; 10; 10

Step 2:
10; 10; 10; 10; 2; 4; 3; 6; 2; 7; 1; 2; 2; 4; 5; 3; 10; 4; 3; 10

Step 3:
18; 12; 18; 4; 2; 5; 9; 16; 1

Test 40

Step 1:
10; 13; 13; 17; 14; 7; 11; 12

Step 2:
10; 14; 12; 12; 4; 13; 12; 13; 4; 11; 10; 9; 10; 10; 10; 16; 12; 12; 6; 10

Step 3:
is less than; equals; is less than; is greater than; is greater than; is less than; is less than; is less than; equals